Amy Samuels

Follow That Fin!

Studying Dolphin Behavior

RAINTREE
STECK-VAUGHN
PUBLISHERS
A Steck-Vaughn Company

Austin, Texas

www.steck-vaughn.com

For my Baiji

Steck-Vaughn Company

First published 2000 by Raintree Steck-Vaughn Publishers,
an imprint of Steck-Vaughn Company.

Copyright © 2000 Turnstone Publishing Group, Inc.
Copyright © 2000, text, by Amy Samuels and Chicago Zoological Society

Library of Congress Cataloging-in-Publication Data

Samuels, Amy.
    Follow that fin!: studying dolphin behavior / Amy Samuels.
  p. cm.—(Turnstone ocean pilot book)
  Includes bibliographical references and index.
    Summary: Follows two biologists as they study the behavior and everyday life of bottlenose dolphins in
Shark Bay, Australia.
  ISBN 0-7398-1230-0 (hardcover) ISBN 0-7398-1231-9 (softcover)
    1. Bottlenose dolphin—Behavior—Juvenile literature. 2. Bottlenose dolphin—Behavior—Research—
Australia—Shark Bay (W.A.)—Juvenile literature. [1. Bottlenose dolphin. 2. Dolphins.]
I. Title. II. Series.
QL737.C432S35 1999   99-14563
559.53'315—dc21   CIP

For information about this and other Turnstone reference books and educational materials, visit
Turnstone Publishing Group on the World Wide Web at http://www.turnstonepub.com.

Photo credits listed on page 48 constitute part of this copyright page.

Printed and bound in the United States of America.

1 2 3 4 5 6 7 8 9 0 LB 04 03 02 01 00 99

Contents

1 A Day with Dolphins

Cindy and I study bottlenose dolphins from our small boat, Circus Dinghy.

I wake up as the sun rises and go outside to check the weather. A week of strong winds has kept us on land. But as I look out over Shark Bay in Western Australia, I can see that this morning is clear and there's hardly a breeze. I rush back inside the trailer to tell my fellow scientist, Cindy Flaherty, the good news. Today we're going to sea to look for dolphins!

Cindy and I get everything ready for the long day to come. We gather together notebooks, pencils, two tape recorders, a camera, lots of film, and two pairs of binoculars. We'll need all this to do our work. It's important that the equipment stays dry, so everything goes into a big cooler. We'll be out all day, so we pack lots of drinking water, sunscreen, and peanut-butter-and-banana sandwiches for lunch. We won't be seeing many other boats while at sea, so a radio is important in case of an emergency. To be safe, we also take along oars and two life preservers.

We carry everything to the beach where our boat, *Circus Dinghy,* is tied up, and we stow all the gear on the boat. Cindy checks the boat's fuel tank to see if there's enough gasoline, while I hook up a device for measuring how deep the water is. Finally we're ready to go.

We leave early in the morning, while the sea is calm. If the water is too rough, it's hard to see the dolphins. As soon as the boat leaves the beach, we begin looking in every direction. We steer the boat through the sea-grass beds to a deep water channel, a path through the bay. Then, we guide the boat into shallow places, always looking. Finally a dark shape slides slowly up out of the water. It's a fin. Suddenly there are three more fins. We've found what we've been looking for—dolphins!

Cindy cautiously steers the boat over to the dolphins, and we slowly follow along behind them. As the boat moves along, we grab our binoculars to get a good look. I open the cooler and pull out our notebooks, pencils, and the camera. We're ready to begin our day with the dolphins.

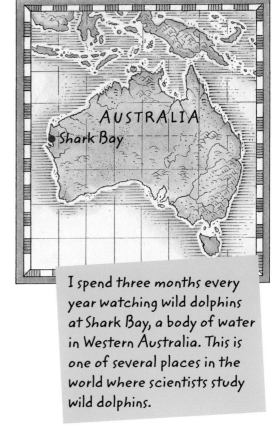

AUSTRALIA
Shark Bay

I spend three months every year watching wild dolphins at Shark Bay, a body of water in Western Australia. This is one of several places in the world where scientists study wild dolphins.

As I follow dolphins, I sometimes use binoculars. I also talk into a tape recorder, describing what I see. Each night I enter into my computer what I recorded during the day.

What Is a Bottlenose Dolphin?

There are more than twenty types of dolphins, including spotted dolphins, spinner dolphins, and bottlenose dolphins. But when people say "dolphin," they are usually thinking of the bottlenose dolphin. Bottlenose dolphins are found in oceans all over the world. Bottlenose dolphins are the stars of movies and the acrobats of aquarium shows. The role of "Flipper" was played by bottlenose dolphins.

Bottlenose dolphins are different sizes in different places around the world. The slender dolphins of Shark Bay are only about 2 meters (6 1/2 feet) long. The dolphins of Moray Firth in Scotland are nearly twice that long, up to almost 4 meters (about 13 feet). Bottlenose dolphins can live for many years, and females often live longer than males. The oldest bottlenose dolphins we know about are more than fifty years old.

Bottlenose dolphins are a kind of ocean mammal called cetaceans (pronounced "se-TAY-shuns"). Cetaceans come in many shapes and sizes, and they swim in all the world's oceans. They include whales, dolphins, and porpoises. Cetaceans share many similarities with mammals that live on land. Cetaceans are warm-blooded, breathe air, and give birth to live young that they feed with milk. But everything cetaceans do—resting, feeding, mating, and raising their young—takes place under the ocean's waves.

Bottlenose dolphins' bodies are smooth and torpedo-shaped to get them through the water easily. ●

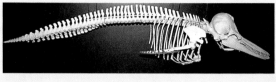

The dolphin's skeleton is long and narrow, an ideal shape for swimming. There are no bones in the dorsal fin or in the tail flukes.

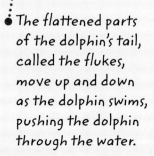

● The flattened parts of the dolphin's tail, called the flukes, move up and down as the dolphin swims, pushing the dolphin through the water.

6

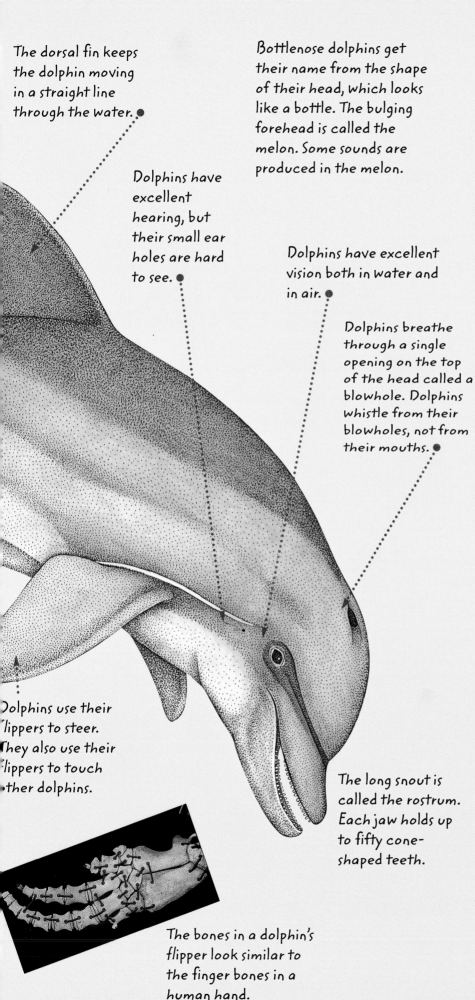

The dorsal fin keeps the dolphin moving in a straight line through the water.

Bottlenose dolphins get their name from the shape of their head, which looks like a bottle. The bulging forehead is called the melon. Some sounds are produced in the melon.

Dolphins have excellent hearing, but their small ear holes are hard to see.

Dolphins have excellent vision both in water and in air.

Dolphins breathe through a single opening on the top of the head called a blowhole. Dolphins whistle from their blowholes, not from their mouths.

Dolphins use their flippers to steer. They also use their flippers to touch other dolphins.

The long snout is called the rostrum. Each jaw holds up to fifty cone-shaped teeth.

The bones in a dolphin's flipper look similar to the finger bones in a human hand.

Spotted dolphins are found in tropical and warm oceans.

Risso's dolphins don't have the pointed snouts many dolphin species do.

Dusky dolphins are often seen in the southern oceans.

The baiji is a rare dolphin that lives in China's Yangtze River.

Taking Attendance

We always start by "taking attendance." We list the dolphins we see, where we find them, and the time we see them. As a dolphin comes to the surface to breathe, we look carefully at the dorsal fin, the triangular fin on each dolphin's back.

To people who study dolphins, dorsal fins are like familiar faces. Some fins are tall and pointed,

You have to look closely to spot the fins in the water. Once I see a fin, I compare its shape and any notches or nicks in the fin with the drawings on my set of "fin flash cards."

Sometimes we use names that go well together. Yin's mother is Yan.

YIN

Squarelet is Square's daughter. Squarelet's fin has a flat tip and a triangular notch on the back.

SQUARELET

while other fins are short and fat. Many fins have nicks and scratches from sharp objects, fishing lines, or from fights with sharks or other dolphins. We make "fin flash cards" to help us remember all the dolphins living in the bay. We also give the dolphins names so that we can remember them. Sometimes we use names that describe the dolphin's fin. For example, Pointer's fin is curved and pointed.

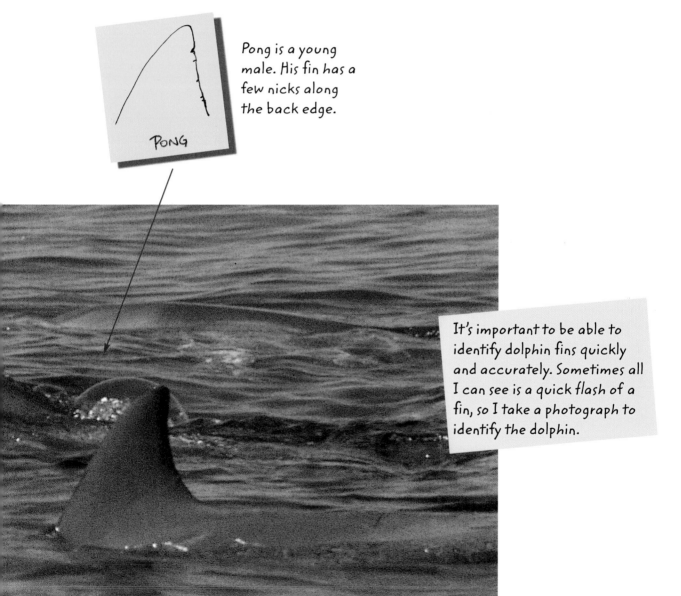

Pong is a young male. His fin has a few nicks along the back edge.

PONG

It's important to be able to identify dolphin fins quickly and accurately. Sometimes all I can see is a quick flash of a fin, so I take a photograph to identify the dolphin.

By taking attendance each day, we can put together a story of each dolphin's life over the years. We know when babies are born and who their mothers, sisters, and brothers are. We know when new dolphins pass through the bay. When a dolphin isn't seen for months, we worry that the dolphin may have died or left the bay.

Looking for Holeykin

Today we recognize two large dolphins right away, without having to look at our flash cards. One dolphin has a broad fin with a jagged tip—that's Nicky. The other dolphin has a curved fin with a ragged edge—that's

Some dolphins—like this male named Captain Hook—have fins that are easy to recognize.

Puck. Nicky and Puck are adult female dolphins. Puck's two daughters, Kiya and Piccolo, are swimming by her side. Kiya is an infant dolphin, or a calf. She's less than one year old and still nurses milk from her mother, just as human babies do. Piccolo is six years old and no longer depends on her mother. She's a juvenile dolphin, a dolphin "kid." I write the dolphins' names in my notebook and take photographs of each one's fin.

We hope to find Holeykin (pronounced "HO-lee-kin"), Nicky's son. Holeykin was named in honor of his grandmother, Holeyfin. At three years old, he is the youngest juvenile in the bay. Holeykin has just stopped nursing milk from his mother and doesn't stay by her side any longer.

Dolphin Family Trees

When very young, a dolphin calf swims close by the mother's side. That is how we know who the calf's mother is and who the sisters and brothers are.

Dolphin fathers do not swim with mothers and calves very often, so we don't know who is the father of each calf.

Here are two dolphin family trees, starting with Square and Holeyfin. The dotted lines connect each dolphin mother to her calves. Researchers did not know Square and Holeyfin when they were calves many years ago, so we don't know when those two dolphins were born.

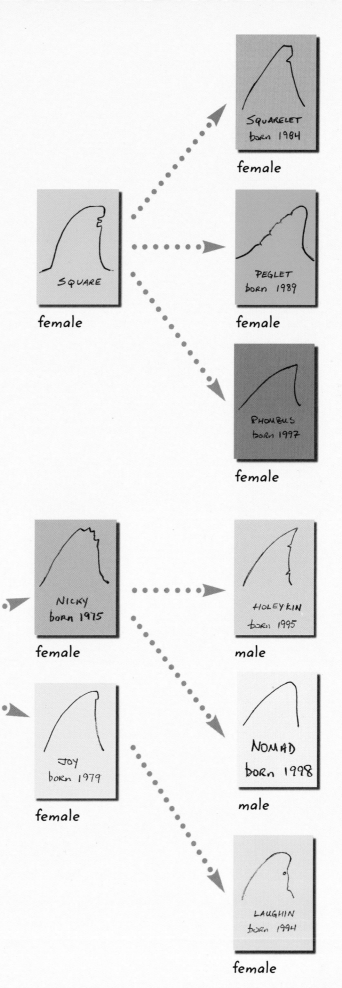

SQUARELET
born 1984

female

SQUARE

female

PEGLET
born 1989

female

RHOMBUS
born 1997

female

HOLEYFIN

female

NICKY
born 1975

female

HOLEYKIN
born 1995

male

JOY
born 1979

female

NOMAD
born 1998

male

LAUGHIN
born 1994

female

Holeykin, the dolphin swimming belly-up, sometimes swims with other juveniles. Here, he's with Smokey, a juvenile male, and Skruff and Laughin, two juvenile females.

We're curious to know how Holeykin is doing, now that he is taking care of himself. Does Holeykin know how to hunt for fish? He seems so small to be on his own. Where is he now?

While we are watching, Nicky, Puck, and Piccolo stop swimming and rest side by side at the surface. They turn back in the direction they just came from. Suddenly we see another dolphin. It's Holeykin speeding closer. He slips into the group, coming to a stop next to Piccolo. He rubs his head against Piccolo's flipper.

We're happy to see Holeykin and add him to our attendance sheet. We spend the next three hours slowly following Holeykin in our boat. Every five minutes, I use my tape recorder to describe where he is, what he's doing, and which dolphins are nearby. I also photograph each dolphin he meets.

Holeykin Notes

After our day at sea, Cindy and I have many recordings and notes to go over before we sleep. We have lots of information about Holeykin stored on a cassette tape. We listen to the tape and write down what we saw during the day. We use shorthand to write the notes quickly. Later, we enter the notes into a computer. Here are some of my notes from watching Holeykin in the morning.

TIME	WHO HOLEYKIN IS WITH	WHAT HOLEYKIN IS DOING		TRANSLATION OF THE SHORTHAND
9:05	NIC, PUC, KIY, PIC	HOL	QS	Holeykin (HOL) swims quickly (QS: "quick-swim") to join Nicky (NIC), Puck (PUC), Kiya (KIY), and Piccolo (PIC).
9:10	NIC, PUC, KIY, PIC	HOL/PIC	RB	Holeykin rubs against Piccolo's flipper (HOL/PIC: Holeykin to Piccolo; RB: "rubs against a flipper").
9:15	NIC, PUC, KIY, PIC	HOL	SS	Holeykin swims alone (SS: "solo-swim"), following the other dolphins.
9:20	NIC, PUC, KIY, PIC	HOL + NIC	ST	Holeykin and Nicky swim together (HOL + NIC: Holeykin and Nicky; ST: "swim together").
9:25	NIC, PUC, KIY, PIC	KIY/HOL	BR	Kiya belly-flops onto Holeykin (KIY/HOL: Kiya to Holeykin; BR: "breaches," or leaps into the air and belly-flops).
9:30	Alone	HOL	SN	Holeykin leaves the other dolphins. He swims belly-up, chasing small fish (SN: "snacking").

Cindy enters her notes into a computer.

We watch as Holeykin trails behind the other dolphins, then swims close by his mother's side. Holeykin plays with the calf, Kiya, who leaps into the air and belly-flops onto Holeykin. Then, Holeykin swims away from the other dolphins to hunt fish to eat. Swimming upside down, Holeykin chases fish at the surface, a type of hunting called "snacking."

Finding Peglet

In the afternoon we search for another dolphin, Peglet. After looking for an hour, we see several fins. Peglet is the first dolphin we

Here, a juvenile male dolphin named Cookie is "snacking." He swims belly-up, chasing a small, silver fish at the surface of the water.

Then suddenly he opens his mouth, whirls around, ...

and he catches the fish.

recognize. Her fin has scars from being tangled in fishing line when she was a calf. Peglet is a juvenile like Holeykin, but she is older and has been on her own for several years. She is now eight years old, but won't be considered an adult for about five more years. Peglet's mother, Square, and Square's new calf, Rhombus, are here, too. Square has a square-shaped fin with two notches on the edge. Rhombus has the smooth, unmarked fin of a calf. I take photographs of each dolphin's fin and write each dolphin's name in my notebook.

We spend the next three hours slowly following Peglet in our boat. I speak into my tape recorder, describing what Peglet does, who she is with, and where she goes. I also photograph each dolphin she meets. We watch while Peglet and her family rest quietly in a tight group. Then, another dolphin arrives. It's Fatfin, an adult female who likes to swim with Square. We recognize her by her fat fin, of course! Fatfin joins the resting group of dolphins.

Family Portrait

This photo was taken when Peglet (right) was still a calf. In those days Peglet's older sister, Squarelet (bottom left), spent a lot of time with Peglet and their mother, Square (middle). Now that Peglet is older, she often swims with Square and her infant sister, Rhombus (not shown in this photograph).

Compare the fins of Peglet's family with the fin flash cards on page 11. (The dolphin between Square and Squarelet is an unrelated juvenile.)

Dolphins have many ways to catch fish. A dolphin's snout can come in handy to uncover small fish in the sand. We call this behavior "bottom grubbing."

Suddenly Fatfin and Peglet swim away quickly. In the distance we see splashing and dolphins jumping. We follow Peglet and Fatfin as they join the dolphins that are feeding on fish. Peglet and Fatfin leap over and over again, chasing and diving after fish. This is a special kind of hunting called "leap-feeding."

It's five o'clock, and we decide to leave Peglet and the other dolphins. It's time to go home to the trailer for our own dinner. We hope the sea will be calm again tomorrow so we can have another day on the water watching dolphins.

Shark Bay is home to several hundred wild bottlenose dolphins, including the 15 juveniles that I study. Many other animals also live in Shark Bay, such as sea turtles, dugongs, sea birds, and fish. Right whales and humpback whales sometimes visit the bay.

Dolphins like to eat this fish, called a "long tom." To avoid being eaten, long toms often slide on the water's surface.

Pelicans nest on Pelican Island in Shark Bay, where there are no animals to eat their eggs or chicks.

Dugongs are marine mammals that mate and give birth in Shark Bay. The bay is home to more than 10,000 dugongs.

These small sharks are no threat to dolphins, but the larger tiger sharks eat dolphins, turtles, and dugongs.

Thousands of loggerhead turtles lay their eggs on the beaches of Dirk Hartog Island in Shark Bay.

2 Studying Animals

I am a type of scientist called a behavioral biologist. Behavioral biologists study what animals do and how they behave. Humans are like animals in some ways, but different in other ways. The way animals do things often looks strange to us. Behavioral biologists are curious about why animals behave the way they do. Many questions about animal behavior haven't been answered yet. Behavioral biologists are working to find those answers. But it's not always easy. And answering one question always leads to more questions.

There are many different things to learn about animals' lives. Some behavioral biologists study how animals live. These scientists try to find out how animals know the right food to eat or how they avoid being another animal's food. Other behavioral biologists study how animals communicate. How do animals let others know that danger is nearby? How do they signal that they are looking for a mate or that they want to be friendly? I am interested in how animals get along with each other. I want to know how they raise their young and which are close companions. I also want to know which animals fight and which have peaceful ways to settle their problems.

This dolphin mother and daughter are in a threatening face-to-face position. Dolphins usually resolve their differences without fighting.

New Ways to Study Animals

How do scientists answer questions about animals? The methods scientists use to study animals have changed over time. Fifty years ago scientists followed a group of animals and tried to write down everything they saw. But this method had problems. The scientists naturally paid more attention to animals that were louder or more active. Because scientists wrote only about animals that caught their attention, sometimes they received wrong answers to the questions they asked.

Some scientists study how dolphins find their food. Here, common dolphins (a type of dolphin) herd, or gather, a school of mackerel into a "bait ball," or tight bunch, and then eat them.

How Do Scientists Watch Animals?

Imagine you are watching your classmates playing outside at recess. You'll probably notice kids who are running or yelling. It's likely you won't pay much attention to a few kids sitting on a bench talking quietly. It's natural that you would notice those who do something to catch your eye or ear. It was the same when scientists looked at a group of animals using old study methods. They noticed the loudest, most active members of the group. They were less likely to notice the quieter members of the group.

For example, forty years ago behavioral biologists who studied baboons thought that male baboons were the leaders and females were the followers. When scientists studied baboon groups, the large, noisy male baboons caught their attention. They didn't watch the smaller baboons, like the mother and baby above, as those baboons sat quietly, calmly grooming each other or napping.

The old methods of studying animals came up with the wrong answer to the question of who are the leaders in a baboon group. The truth is that most of the baboons in a group are females and their babies. The females are related to each other and live together all their lives. Males stay in a group only for a short time. Using new study methods, scientists discovered that even though the males are larger than the females, it is usually the female baboons who decide where the group goes each day.

I used the new methods to study baboons in southern Kenya in East Africa.

20

Heidi Notes

TIME	WHO IS HEIDI NEAR?	WHAT IS HEIDI DOING?	NOTES ABOUT HEIDI
7:15	Juveniles Omo, Dudu, and Aloe	Playing	Baboon kids play in the sleeping trees while the adults wake up.
7:30	Mother Handle & baby sister Haki	Grooming	Heidi picks through her mother's fur and removes dirt.
7:45	Brothers Hans & Hodi (pronounced "HO-dee")	Walking	The baboons are hungry. Heidi follows her big brothers as the group crosses an open plain looking for food.
8:00	Handle & Haki	Walking	Heidi stays near her mother while they walk through tall grass where lions might be hiding.
8:15	Hodi	Feeding	Heidi eats ripe berries with her big brother.
8:30	A big male, Nol	Avoiding	Nol pushes Heidi away from the berries, and she runs away. Nol eats the ripe berries.
8:45	No one	Feeding	Heidi eats berries by herself. She is never really alone—she can hear baboons eating in the bushes all around her.
9:00	Omo	Resting	After eating, the baboons take naps in the shade near a water hole. Zebras and giraffes come by for a drink.

Now scientists use new methods to observe animals. Each animal in a group is watched for the same amount of time. A scientist watches one individual in a slow, careful way. The scientist concentrates for a long time on what the animal is doing, even if the animal is only napping. It's important to know about everyday activities in animals' lives. Also, at any moment an animal may do something really amazing, and the scientist wants to be watching when that happens.

Here are my notes from watching a juvenile female baboon, Heidi, one morning. Scientists can take detailed notes like these, or they can mark certain behaviors on a checklist. Using notes or a behavior checklist, scientists then count how often an animal does a behavior or which animals do which behaviors.

Animal Societies

Many female mammals have close relationships with their female relatives. In groups of baboons, monkeys, elephants, hyenas, pilot whales, sperm whales, and bottlenose dolphins, female relatives all live in the same community for their entire lives. Animal societies based on female family ties are called matrilineal societies.

It's unusual for male mammals to have close relationships with other males. It is more common for males to avoid or fight with each other. Close relationships among males are found only among a few kinds of mammals. A male chimpanzee may have one or a few other males as his close companions. Sometimes these are his relatives. Scientists don't know if male dolphins that are close companions are related.

Many kinds of mammals live in matrilineal societies. Female African elephants and female Indian bonnet monkeys live with their female relatives for their entire lives.

Lion brothers help each other find mates and fight other males.

In the past scientists tried to write down each and every thing that an animal did, from an ear twitch to a leap. Today scientists often use a checklist of behaviors that interest them. The careful records, or data, that scientists gather from watching animals are entered into a computer. Counts of the data reveal interesting things about animals' behavior. Scientists then use graphs and charts to illustrate what they have learned.

The new methods were first used to study the behavior of primates, which are animals such as chimpanzees and baboons. Dolphins were usually studied as a group using the old methods. When I first became a scientist, I studied baboons and learned to use the new methods. When I became interested in dolphins, I tried using the new methods to study dolphins. I'm one of the first scientists to study dolphins in this new way.

I used the data I collected while observing several juvenile dolphins to create this graph. I wanted to know what percentage of the total observation time the dolphins spent near their mothers. The graph shows that juvenile daughters stayed near their mothers more than juvenile sons did. Some of the juvenile sons weren't seen near their mothers at all.

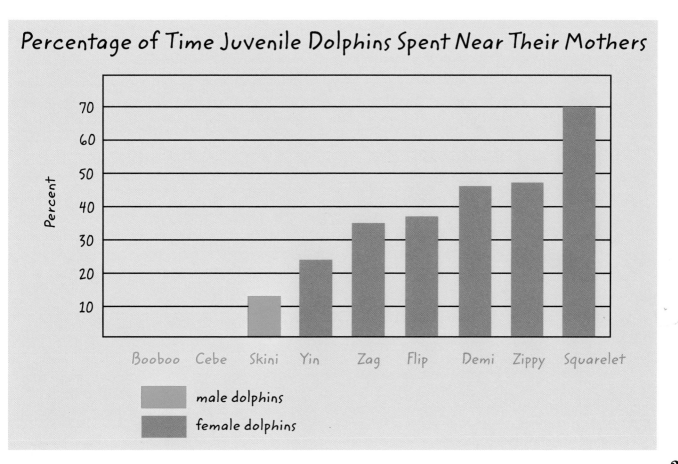

Percentage of Time Juvenile Dolphins Spent Near Their Mothers

I follow dolphins day after day because I want to learn about juvenile bottlenose dolphins—dolphins that aren't grown up yet, but aren't babies, either. Like human children, dolphin "kids" have many things to learn about the world in which they live. I want to know how juvenile dolphins learn how to hunt for fish and how to avoid dangers such as sharks and boats. I follow each juvenile for several hours at a time and record the data.

Other Behavioral Biologists Who Study Dolphins

Another behavioral biologist, Richard Connor, studies adult male bottlenose dolphins in Shark Bay. Richard uses the new study methods, too. In a small boat, he follows male dolphins to learn about

These three adult male dolphins are close companions. They often swim close together and come to the water's surface at exactly the same moment.

their behavior. He has learned that adult male dolphins often have other males as their close companions. These males swim together in small groups. Richard wants to know why adult male dolphins form close attachments with other males.

Adult male dolphins do not appear to form close attachments with adult females. Males and females come together to mate, and after mating they go their separate ways. Richard wants to know which males mate with which females and which male is the father of each calf.

Male dolphins that are close companions sometimes swim on either side of an adult female when it's time to mate.

Peter Tyack listens to the underwater sounds of dolphins. He wants to know what the whistles and clicks mean.

Peter Tyack is a behavioral biologist who studies the sounds bottlenose dolphins make. Dolphins make clicking sounds to find fish and whistling sounds to communicate with each other. Peter thinks that dolphins may recognize each other by those whistles. Researchers think that each dolphin has a unique whistle, called a signature whistle. Does the signature whistle let others know where the dolphin is?

Peter is working to answer this question by studying dolphins in the ocean and in aquariums. He uses a hydrophone, an underwater microphone, to listen to the dolphins' sounds, and he watches to see what dolphins do when they hear the signature whistle of a distant dolphin.

Many behavioral biologists are studying how dolphins live and interact. Together they are working to create a better picture of dolphins' lives.

Sometimes you can see a stream of bubbles coming from the blowhole when a dolphin whistles. Then it's easy to tell which dolphin is making a sound. But usually there is no visible sign to show which dolphin is whistling, so it can be difficult to study dolphin sounds.

Listening to Dolphins

It can be hard to study dolphin sounds. It is often difficult for a scientist to tell which animal is making a sound under the water. When people talk, their lips move, but when dolphins make sounds, there are usually no obvious signs. More than ten years ago, Peter found a way to solve this problem. He invented a device called a vocalight. Using vocalights, Peter could tell which dolphin was making the sound.

Now there are more advanced devices to record dolphin sounds. For example, a tiny tape recorder can be attached to a dolphin's dorsal fin with suction cups. After several hours the recorder falls off and is retrieved by the scientist. When the tape is played into a computer, the sound is translated into a picture called a spectrogram.

At right are spectrograms of the signature whistles of four different dolphins from Florida. The numbers along the bottom of each spectrogram show the amount of time in seconds. The numbers along the side of each spectrogram measure each whistle's frequency, or how high or low the sound is. Frequency is measured in kilohertz (kHz). The higher the number, the higher the sound.

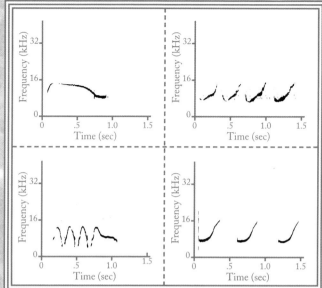

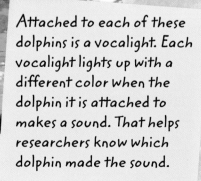

Attached to each of these dolphins is a vocalight. Each vocalight lights up with a different color when the dolphin it is attached to makes a sound. That helps researchers know which dolphin made the sound.

3 A Close-up View of Dolphins

A juvenile female often swims with her mother. In this photograph Piccolo (top) is swimming with Puck, her mother, and Kiya, her baby sister. Perhaps this is how juvenile female dolphins learn how to be mothers.

Scientists can't assume that other animals think, see, or feel things the same ways we humans do. Animals' lives are mysteries for behavioral biologists to solve. So learning about animals is like detective work. To find out what a dolphin's underwater world is like, scientists must look for clues. To do that, scientists travel to places where they'll find dolphins. They watch what dolphins do and take careful notes about what they see. Then, they try to

fit the bits of evidence together like the pieces of a puzzle. It often takes years of careful work to solve animal mysteries.

To find pieces of the puzzle, I study individual dolphins for months or even for years. After watching the dolphins for so much time, I get to know each one very well. Every dolphin has a distinct personality, just like people do—some are relaxed, some are careful, and some are bossy.

I also learn to see patterns in the dolphins' behavior. Dolphins have different personalities, but some dolphins share certain ways of doing things. After watching the dolphins closely, I learn which dolphins are similar. I learn the special behaviors of adult males, how mothers with calves act, and what things the "kids" like to do.

Juvenile males like Holeykin do not spend much time with their mothers or with calves. Instead, they wrestle and play with other young males. Perhaps this is how each juvenile male chooses the male who will be his close companion when he grows up.

We often follow the dolphins at a distance, and we try not to disturb them. Sometimes we need to use binoculars to see what the dolphins are doing.

When I am in Australia, I wake up each morning wanting to know which dolphins I will see during the day. Will I see Piccolo, Holeykin, and Peglet? Who will each one be with? What will they be doing? But finding a small dolphin like Holeykin in a big area of ocean can be difficult. A glimpse of a fin is usually the only clue to where a dolphin might be. Even after we find dolphins, it can be hard to see what they're doing because they're underwater most of the time.

Scientists can do two things to learn more about dolphins. They can go where the dolphins live in the wild, or they can go to see dolphins in zoos and aquariums.

When I go to where the dolphins are, I go to Shark Bay. I follow wild dolphins in a small boat, moving slowly so that the boat won't disturb the dolphins. I never go in the water to watch dolphins because I don't want to interrupt what the dolphins are doing. Also, it can be a dangerous place to swim. There is a reason it's called Shark Bay—it's full of sharks!

But observing from the boat works well. The water in the bay is clear and not very deep. When I peer over the side of the boat, I can usually see what the dolphins are doing underwater.

There are many threats to wild dolphins' lives. The threats include a shortage of fish to eat, illness, and shark attacks. This dolphin's back shows the scar from a shark attack.

Life in a Bottlenose Dolphin Community

There may be a hundred or more individual dolphins in a bottlenose dolphin community. Bottlenose dolphins do not travel together in one large group. Instead, the dolphins move from one small group to another, visiting many other dolphins each day. Some dolphins form strong attachments to each other and are often seen swimming together.

After a pregnancy of one year, a dolphin mother gives birth to a single calf. Calves are usually born in the summer months. A calf stays with his or her mother, drinking the mother's milk, for two to five years. The bond between a mother and her calf is the closest relationship in a dolphin community. A mother and calf swim close together, as in the photograph at top left.

After a calf stops drinking the mother's milk, the dolphin is considered a juvenile. A juvenile female becomes an adult and can have her first calf when she is between 6 and 13 years old. A male becomes an adult and can father his first calf when he is between 8 and 15 years old.

Adult male dolphins form bonds that are nearly as close as the bond between a mother dolphin and her calf. An adult male usually has one or two other males as his close companions. Male dolphin companions may stay together for many years.

Adult female dolphins have many female companions. A mother and her calf will often swim together with other mothers and calves. But the closest companions of adult females are their female relatives. Mothers, daughters, sisters, and aunts visit each other often. A dolphin community is made of several groups of female relatives.

The other way I study dolphins is at a zoo. I work at Brookfield Zoo, near Chicago, and I study dolphins there. There are special windows built into the walls of the dolphin pool. Looking through the window, I can see everything the dolphins are doing underwater. I can see dolphins gently touch each other with a flipper or a snout, or a calf's tongue curl while nursing from the mother dolphin. I can see who picks a fight and who wins. The underwater window is my "spy hole."

Watching dolphins from my spy hole at the zoo and from a boat in the wild are both good ways to find clues to dolphin-behavior mysteries. One mystery I'm interested in is a behavior called "gentle rubbing." I have learned that dolphins who like to swim together also like to take turns rubbing each other. But why do they do this?

I gather information here at the zoo and in Australia. When I'm not watching dolphins, I study my notes and photographs at the Woods Hole Oceanographic Institution in Massachusetts.

Dolphins often take turns rubbing each other. Here, two adult males at Brookfield Zoo, Nemo and Stormy, swim side by side. Stormy holds out a flipper for Nemo to rub against.

Stormy

Nemo

Yellow baboons in southern Kenya groom each other. Grooming is a way that monkeys show close relationships and keep their hair clean.

I can tell rubbing is a friendly action because the dolphins touch each other so gently. But I think there may be another reason.

When I look closely at dolphins rubbing, I see something else. Tiny bits of skin come off when the dolphins rub each other. This gives me the idea that gentle rubbing may be a way for dolphins to keep their skin smooth and clean. Like monkeys that pick through each other's hair with their fingers, dolphins seem to be grooming each other with their flippers. More observations of dolphins at the zoo and in the wild may show whether my idea is right.

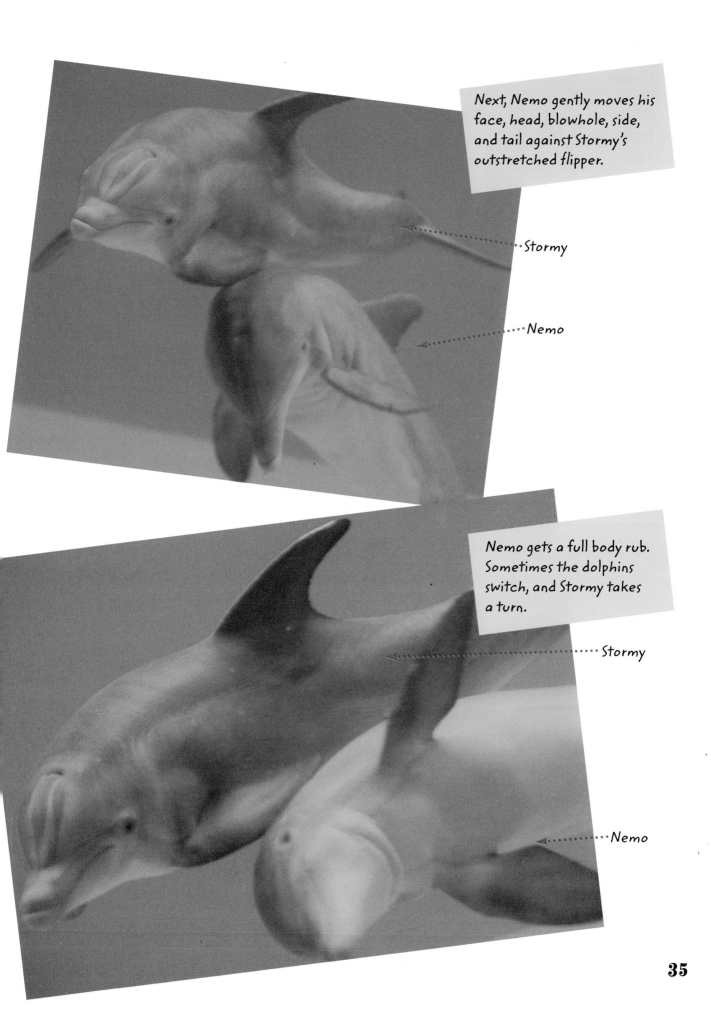

Next, Nemo gently moves his face, head, blowhole, side, and tail against Stormy's outstretched flipper.

···Stormy

···Nemo

Nemo gets a full body rub. Sometimes the dolphins switch, and Stormy takes a turn.

······Stormy

···Nemo

Studying Ocean Animals

How do humans, who live on land, study animals that live in the sea? Scientists have found many solutions to this problem. In Argentina scientists watch from the tops of cliffs that look out over a bay where mother right whales care for their calves. In California scientists watch gray whales from cliffs as the whales migrate between the Arctic and Mexico. In the Arctic scientists in airplanes follow bowhead whales far out to sea. In the Caribbean and in other places, scientists study sperm whales by listening to their sounds through hydrophones.

In Hawaii scientists built a special boat to study spinner dolphins. Under the boat there's a small room with windows. It is just large enough for one person to sit. In the Bahamas scientists carry a video camera while they swim with spotted dolphins.

Each of these methods helps solve the problem of how to study ocean animals. But each method has its own problems. Watching from a distance can only give scientists an overall view of the animals. Listening underwater with a microphone, scientists can hear the clicks and creaks of sperm whales, but it's difficult to tell which whale is making which sound.

An underwater viewing compartment provides an amazing view of the world under the waves. But the rocking and rolling of the boat can make a person sitting in the compartment terribly seasick. A human swimmer is much slower than a dolphin, so a scientist in the water can only catch a glimpse of dolphins as they speed by.

Scientists attach a video camera to a helium balloon and tow the balloon behind a boat. This gives them a bird's-eye view of dolphins hunting for fish.

Studying dolphins in the ocean and at a zoo provides clues about another dolphin mystery. One day at the zoo, two adult males began moving together in a display that looked like a dance. They were swimming so closely to each other that they nearly touched. Each dive to the bottom of the pool and every breath of air at the surface were done at exactly the same time. It was as if they were a single dolphin.

They leaped into the air, slapped their tails on the water, and wiggled their flippers back and forth perfectly together. They even spit water at the same time, like two fountains. Why were they doing this? What kind of display was this?

Many months later I learned part of the answer to the mystery from Richard Connor. He had seen male dolphins in the wild do something similar, a behavior he calls a "synchronous display."

At the zoo two males do a synchronous display. They swim side by side, moving their tails and flippers in exactly the same way.

Male dolphins in Shark Bay also do synchronous displays. Here, two males belly-flop at the same time on either side of a female.

Richard discovered that this display is done by male dolphins who are close companions. So we know who does the display, but we don't know why they do it.

The display may be a warning to other male dolphins, signaling, "We're a team. Look out!" Or it may be a way for males to attract female dolphins. We are now looking for more clues to the mystery. This is the way scientists solve animal mysteries. We start with questions and then work to find the answers, clue by clue. And there are always more questions to answer.

Here, two juvenile males in Shark Bay swim side by side and come to the surface at exactly the same moment, just like adult male dolphins do. Are they just playing, or are they learning to behave like adults?

Dolphins at the Zoo and at Sea

There are some differences in the lives of dolphins at a zoo and in the wild, but they also have many things in common. Wild dolphins have to work hard to survive. They spend much of each day searching for fish to eat. Dolphins at a zoo learn to work with their human caretakers in exchange for plenty of fish. Zoo dolphins don't have to spend as much time searching for food as wild dolphins do, so zoo dolphins can spend more time resting and visiting with each other than wild dolphins can.

Zoo dolphins and wild dolphins are alike in some ways. The ways they touch each other look the same whether the dolphins live at the zoo or in the ocean. Both zoo and wild dolphins gently rub each other. Zoo and wild dolphins also fight in the same way—biting, or hitting with their snouts, dorsal fins, or tails.

Zoo and wild dolphins also swim touching each other, with one dolphin's flipper resting against the other dolphin's side. That way, one dolphin takes a ride. This behavior is called "contact swimming."

4 Caring About Dolphins

Nicky brings her
newborn calf, Nomad,
to the beach at
Monkey Mia.

I am also a conservation biologist because some of the research that I do helps protect the dolphins that I study. I want to know whether interactions with humans are harmful to wild dolphins, so I study Nicky and Puck, Shark Bay dolphins that have an unusual life. They visit a beach called Monkey Mia (pronounced "MY-ah"). People travel from all over the world to meet them.

For Nicky and Puck, visiting the beach is a dolphin family tradition. Nicky's mother, Holeyfin, and Puck's mother, Crooked Fin, both came to Monkey Mia nearly every day of their lives.

At Monkey Mia the park rangers give Nicky and Puck a few fish to eat. Since scientists have found that many wild animals are harmed when people give them food, rangers carefully control how much food they give the dolphins. Each dolphin is given only a small amount of fish each day so the dolphins must continue hunting for themselves. That way, Nicky and Puck will continue to be wild dolphins.

Roxane Shadbolt, a park ranger at Monkey Mia, is feeding Nicky a few fish.

Watching the dolphins at Monkey Mia is a very popular pastime for the area's visitors.

Puck's daughter, Piccolo, will soon be offered fish by the rangers to continue her dolphin family tradition. But before that happens, Cindy and I will watch Piccolo very carefully to make sure that she has learned how to hunt fish for herself. We don't want her to depend on the fish that the rangers give her. We want Piccolo to stay a wild dolphin.

Some dolphins in other parts of the world are not so lucky. In Florida and South Carolina, some dolphins are given fish by people in boats, even though it is illegal in the United States. In those places no rangers are protecting the dolphins. Now I am studying the dolphins in Florida to find out if they are in danger. Juvenile dolphins fed by people may not learn how to hunt for themselves. Dolphins that come close to a boat to beg for fish can be hurt by the boat's propeller. Dolphins that come close to fishing gear to look for fish can become tangled in fishing line or can accidentally swallow a hook.

People who feed wild dolphins don't understand that they may be putting the dolphins in danger. I hope that my research in Florida will help protect the dolphins living there.

Animal research can also help zoos and aquariums take care of their dolphins. At Brookfield Zoo, Cindy and I watch each calf that is born to make sure the baby is healthy and looked after by the mother. We tell the

Piccolo has never been fed by the rangers, but she's been watching while Puck, her mother, is fed. When she comes to the beach with Puck, Piccolo catches her own fish and offers it to the rangers. Here, Piccolo is bringing a fish to "feed" the rangers.

zookeepers when one dolphin swims alone too much because that dolphin may not be feeling well. From the studies we do at sea and at the zoo, we know that related females and pairs of males like to stay together. These studies help the zoo decide which dolphins will live together peacefully.

By watching dolphins in the wild and at zoos, scientists are beginning to uncover some of the secrets of the dolphins' hidden world. But there is always much more to learn. I want to find out why dolphins rub each other. What is it like to be a dolphin "kid"? Why do male dolphins do a display that looks like a dance? Are dolphins harmed when people swim with them in the ocean? I've learned a lot about the lives of the dolphins, but the more I learn, the more I want to know.

Windy swims with her newborn calf, Kai, at Brookfield Zoo. Baby dolphins are born with stripes on their sides. The stripes remain there for several months.

Glossary

behavioral biologist A behavioral biologist is a scientist who studies what animals do and how they act.

blowhole A bottlenose dolphin breathes air through a single hole in the top of the head, called a blowhole. A dolphin produces whistles using the blowhole, not the mouth.

bottom grubbing Bottom grubbing is a way that bottlenose dolphins hunt for fish by using their snouts to uncover fish that are hiding under the sand.

calf An infant dolphin is called a calf. A calf stays with the mother, drinking the mother's milk, for two to five years.

cetacean [se-TAY-shun] A cetacean is a type of mammal that lives in the water. Cetaceans include whales, dolphins, and porpoises.

conservation biologist A conservation biologist is a scientist who studies animals and plants to protect them.

contact swimming Contact swimming is a friendly behavior in which two dolphins swim while touching each other.

data The careful records that scientists gather are called data. A single scientific record is called a datum.

dorsal fin The triangular fin on a dolphin's back is called a dorsal fin. This fin helps the dolphin to swim in a straight line through the water. Scientists look at nicks, notches, and the shape of the dorsal fin to recognize individual dolphins.

flipper Dolphins have a flipper on either side of the body to steer through the water and to touch other dolphins. The bones inside look like the fingers of a human hand.

flukes Flukes are the flattened part of the tail of an animal such as a dolphin or whale. The flukes move up and down to push the animal through the water.

gentle rubbing Gentle rubbing is a friendly behavior in which one dolphin holds out a flipper for another dolphin to rub against. Gentle rubbing may also be a way that dolphins keep each other clean.

grooming Grooming is a friendly behavior and a way that animals like baboons keep their bodies clean.

juvenile A juvenile dolphin is older than a calf, but not yet grown up—a dolphin "kid." A juvenile female becomes an adult and can have her first calf when she is between 6 and 13 years old. A juvenile male becomes an adult and can father his first calf when he is between 8 and 15 years old.

leap-feeding Leap-feeding is a way that bottlenose dolphins hunt for fish. The dolphin leaps out of the water over and over, each time diving under the water to chase fish.

mammal A mammal is a type of warm-blooded animal that breathes air and gives birth to live young that are fed with milk. Humans and bottlenose dolphins are types of mammals.

matrilineal society A matrilineal society is an animal community that is based on the ties among female family members.

melon The bulging forehead of a bottlenose dolphin is called the melon. Some sounds are produced in the melon.

rostrum [RAH-strum] The snout of a bottlenose dolphin is called the rostrum. A dolphin's rostrum can be used to dig in the sand for fish and to hit or touch other dolphins.

signature whistle Scientists believe that each bottlenose dolphin produces a unique whistle called a signature whistle.

snacking Snacking is a way that young bottlenose dolphins hunt for fish. The dolphin chases a small fish while swimming belly-up at the water's surface.

synchronous display Pairs of male bottlenose dolphins sometimes swim together in a display that looks like a dance. The two male dolphins move together as if they were a single dolphin.

Further Reading

Corrigan, Patricia. *Dolphins for Kids.* Minoqua, WI: Northword Press, 1995.

Goodall, Jane. *My Life with the Chimpanzees.* New York: Minstrel Books, 1967.

Houghton, Sue. *Dolphin.* Mahwah, NJ: Troll, 1994.

Mallory, Kenneth. *A Home by the Sea: Protecting Coastal Wildlife.* San Diego, CA: Harcourt Brace, 1998.

Moss, Cynthia. *Little Big Ears: The Story of Ely.* New York: Simon & Schuster, 1997.

Parker, Steve. *Look into Nature: Whales and Dolphins.* San Francisco: Sierra Club Books for Children, 1992.

Pringle, Laurence P. *Dolphin Man: Exploring the World of Dolphins.* New York: Atheneum, 1995.

Reeves, Randall, and Stephen Leatherwood. *The Sea World Book of Dolphins.* San Diego, CA: Harcourt Brace Jovanovich, 1987.

Stoops, Erik D., Jeffrey L. Martin, and Debbie L. Stone. *Whales.* New York: Sterling Publishing, 1995.

Sweeney, Diane, and Michelle Reddy. *Dolphin Babies: Making a Splash.* Boulder, CO: Roberts Rinehart, 1998.

Index

Acknowledgments

My parents, Peggy and Harold Samuels, encouraged me to write for children, and my editors, Audrey Bryant and Debby Kovacs, gave me the opportunity. I am grateful to them all. I thank Sukie Kaufmann, Ali Baker, and Laura Samuels for letting me know what kids want to know about animals. David Charles, Richard Connor, Cynthia Flaherty, Sophia Fox, Doug and Stephanie Nowacek, Roxane Shadbolt, and Peter Tyack made helpful suggestions for the book. The Shark Bay Dolphin Research Project, the Department of Conservation and Land Management of Western Australia, the rangers of Monkey Mia, the Monkey Mia Resort, Monkey Mia Wildlife Sailing, Brookfield Zoo's Seven Seas Panorama, the U.S. Marine Mammal Commission, the U.S. National Marine Fisheries Service, and the Amboseli Baboon Project are gratefully acknowledged for their contributions to the research described in this book. In memory of Akea, Angie, Connie, Finnick, Holeyfin, Shana, Squarelet, and Windy.

Photographs courtesy of Amy Samuels from the Chicago Zoological Society, except for the following:

Altmann, Jeanne: 20 bottom; Bonner, Vicki: 4; Colla, Phillip/Innerspace Visions: 7 second from top; Connor, Richard: 24, 25 bottom, 37 bottom; Flaherty, Cindy/Chicago Zoological Society: 18, 28, 32 top, 38; Fleetham, D./*Animals Animals*: 17 third from top; Gordan, Ethan: 3 bottom; Greer, Mike/Chicago Zoological Society: 3 top, 13, 26 top, 27 bottom, 34 top, 35, 39 bottom, 43; Kleindinst, Tom/Woods Hole Oceanographic Institution (WHOI): 5; McGarrity, Cheryl/Chicago Zoological Society: 39 top; Minakuchi, Hiroya/Innerspace Visions: 7 third from top; Nowacek, Doug/WHOI: 36; Perrine, Doug/Innerspace Visions: 2, 7 bottom left, 7 top right, 16; Raven, Harvey/Monkey Mia Wildlife Sailing: 30; Richards, Andrea: 25 top; Sayigh, Laela: 27 top; Schulz, Jim/Chicago Zoological Society: 6, 33; Seitre, Roland/Innerspace Visions: 19, 7 bottom; Shah, A. and M./*Animals Animals*: 22 bottom; Tyack, Peter: 26 top; Watt, James/*Animals Animals*: 1, 17 fourth from top.

Illustration on page 5 is by David Stevenson.
Illustration on pages 6–7 is by Patricia Wynne.